Traditional Iroquois Corn
Its History, Cultivation, and Use

Jane Mt.Pleasant

Natural Resource, Agriculture, and Engineering Service
Cooperative Extension
PO Box 4557, Ithaca, NY 14852–4557

NRAES–179
March 2011

ISBN 978-1-933395-14-2

Library of Congress Cataloging-in-Publication Data

Mt. Pleasant, Jane, 1950-
Traditional Iroquois corn : its history, cultivation, and use / Jane Mt. Pleasant.
p. cm. -- (NRAES ; 179)
Includes bibliographical references.
ISBN 978-1-933395-14-2
1. Corn. 2. Corn--New York (State)--History. 3. Iroquois Indians--Agriculture--History. 4. Cooking (Corn) I. Natural Resource, Agriculture, and Engineering Service. Cooperative Extension. II. Title. III. Series: NRAES (Series) ; 179.
SB191.M2M74 2011
633.1'509747--dc22

2010031352

Disclaimer

To order additional copies, contact:

Natural Resource, Agriculture, and Engineering Service (NRAES)
Cooperative Extension
PO Box 4557, Ithaca, New York 14852-4557
Phone: (607) 255-7654 • Fax: (607) 254-8770
Email: NRAES@CORNELL.EDU • Web site: WWW.NRAES.ORG

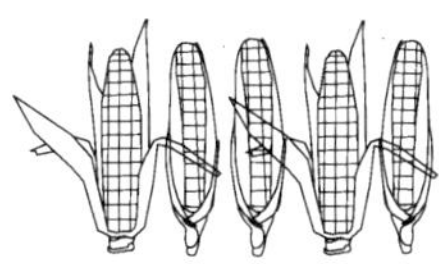

Contents

continued on next page

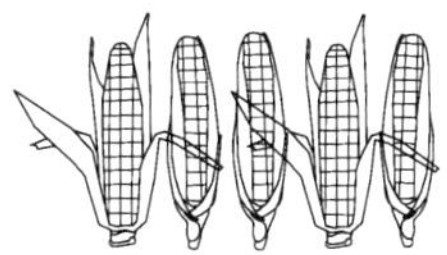

Acknowledgments

I owe many thanks to many people who have shared their knowledge, labor, and passion for Iroquois corn. Here I mention just three who stand large in terms of their contributions to this book. John Mohawk, Seneca intellectual and tireless advocate for traditional Iroquois white corn, inspired me to grow corn and taught me to relish eating it. He not only loved corn, he also grew it when few others did. Norton Rickard, Tuscarora cousin, for decades grew substantial acreages of traditional Iroquois corn on his farm in Tuscarora territory. He provided a bridge of knowledge across generations that ensured the corn's survival into the 21st century. With his family, he nurtured community pride and connection to this traditional food crop, hosting annual husking bees and providing corn for seed and food to Haudenosaunee people near and far. Robert Burt, retired research support specialist at Cornell, and I have worked together for more than twenty years, cultivating Iroquois white corn in many locations, as part of field research, farm demonstrations, and production fields to increase seed supply. Without Bob's extensive knowledge of crop production, and his hard work and commitment, this book would not be possible.

Jane Mt.Pleasant

Thanks to the following peer reviewers whose input enhanced the quality of this book: Julie Brussell, University of New Hampshire Cooperative Extension; John P. Hart, New York State Museum; Frank Kutka, Dickinson Research Extension Center, ND; Klaas and Mary-Howell Martens, farmers and grain dealers, Penn Yan, NY; Audra Simpson, American Indian Program, Cornell University; and Margaret E. Smith, Plant Breeding and Genetics, Cornell University.

The book was edited by Jamie Leonard and designed by Yen Chiang, NRAES.

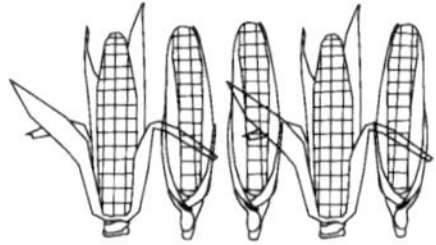

About the Author

Jane Mt.Pleasant is both director of the American Indian Program and associate professor in the Department of Horticulture at Cornell University. Considered a national expert in Iroquois agriculture, Mt.Pleasant's research focuses on indigenous agriculture and its links to contemporary agricultural sustainability.

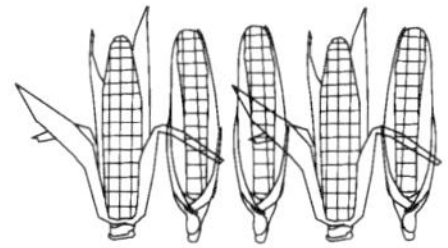

Introduction

Corn is so ordinary and ubiquitous in contemporary life that we often overlook its profound, transformative powers. We drive by mile after mile of cornfields on a summer day, seldom even glancing at this strange, yet incredibly productive plant that has shaped human history in the Western Hemisphere for more than five thousand years. This book provides a more intimate look at corn in Iroquoia, an area that stretches from Ontario, Canada south to the Appalachian Region and west to the Ohio Valley. Although many people think that

Figure 1 **Iroquois Corn in Bushel.** This eight-rowed corn with long slender ears and white kernels is called Iroquois (or Tuscarora) White Flour Corn. It has been grown in Iroquois communities for several hundred years and is still used to make traditional corn soup and corn bread. *Courtesy of Jane Mt.Pleasant.*

the first inhabitants of this area were long ago displaced and disempowered, the Iroquois (also called the Haudenosaunee, People of the Longhouse, or the Five and later Six Nations Confederacy) and their political, social, cultural, and economic institutions still exist and even thrive throughout the region. More than any other single aspect of Iroquois life, corn reflects the persistence of Iroquois people, the complexity of their cultural institutions, and their enormous contributions to contemporary life in the region.

Although all corn varieties cultivated today are descendants of corn that was developed and cultivated by indigenous peoples, this book focuses on the traditional, open-pollinated Iroquois varieties still grown today (see figure 1, page 1), which are similar to those that were being cultivated when European colonizers first landed in northeastern North America. When I plant this traditional Iroquois corn, I feel intimately connected to generations of Iroquoian farmers who have participated in this same activity each spring for centuries. In fact, the only reason I can plant this corn today is because of the careful, persistent, and knowledgeable attention of indigenous farmers who have cultivated corn across varied landscapes and many generations. Corn cannot exist without human intervention. So every spring when I plant corn again, I join that long line, stretching back thousands of years, of corn cultivators who have ensured the survival of this crop for future generations. I hope to share with readers that sense of significance and connection to corn.

This short book will be useful to a wide range of people: those interested in Iroquois history and culture; educators who want a more thorough understanding of Iroquois agriculture and its significance; gardeners or farmers interested in planting open-pollinated varieties of corn; and anyone who wants to expand their culinary experiences with corn beyond corn on the cob and popcorn. But most importantly, I hope that Haudenosaunee people will find the book useful in recovering and reasserting our intimate connections to corn. If I inspire both Native and non-Native peoples to plant open-pollinated varieties of corn, take pleasure in their care and cultivation, and then prepare them in delicious and healthy foods, I will be delighted and will have achieved my aim for writing this short book.

The book begins with an Iroquois understanding of how corn came to be entwined with human life and traces its impact on and within Iroquois history. Corn plays important and recurring roles in three key oral texts that describe 1) the beginnings of life on Turtle Island, 2) Haudenosaunee governance, and 3) Haudenosaunee worldview. I also describe how western academics (ethnobotanists, archeologists, and historians) explain corn's arrival in the Northeast and then explore corn's entanglement with both Haudenosaunee and Euro-American communities from the colonial period through contemporary times. A section on the botany of corn tries to make sense of all the different forms that corn takes and also provides information on corn reproduction, with explanations of terms like "variety," "open-pollinated," and "hybrid."

If you are just interested in growing corn, you will find practical instructions for planting corn by hand in a home garden or in a larger acreage that may entail the use of hand tools or even small machinery. Here you can find a discussion on the pros and cons of tillage, the use of hills, information on when and how many kernels to plant, how to control weeds, and how to maintain soil fertility, whether you use practices that are organic, conventional, or something in between.

Saving seed for planting next year is a surprisingly important activity, and knowing how to do it well ensures that future generations will have these valuable varieties. The section on corn harvest includes information on how to store corn safely in order to minimize damage from molds, disease, insects, and other pests.

The last section on preparing corn provides information on nixtamalization, a process developed by Native peoples that involves cooking corn with wood ash to increase its nutritional value while decreasing the cooking time. This section also includes information on the nutrition of corn and its importance in healthy diets.

I intend this book to serve as a gateway to other sources of information. At the back of the book, you will find additional resources for each of the major sections of the book. These are sources that I used to help write the book so, you can check on the accuracy of my representation of these other experts. But this collection of books, journal articles, Web sites, and organizations also provides avenues for exploring each of the topics in more depth. Whether you are a teacher who wants to better understand the botany of corn, a Haudenosaunee community member who desires more accurate information on the role of corn in history, or an avid gardener interested in growing open-pollinated corn, I hope that each of you will be inspired to connect more deeply to this food crop that has so profoundly shaped our lives.

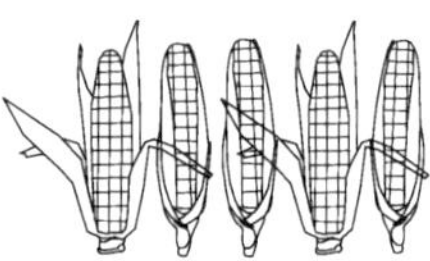

History of Iroquois Corn

The Beginnings of Corn: A Haudenosaunee Perspective

Native Americans have been growing corn in the Northeast for almost two thousand years. By the middle of the fourteenth century, corn served as the major crop for the Haudenosaunee who used it as the foundation of stable and productive agriculture across the region. Corn has been central to Iroquois culture, playing a pivotal role in the social, political, economic, and cultural arenas of indigenous life. It appears in key Iroquois oral texts including the Creation Story, the Great Law of Peace, and the Thanksgiving Address. Throughout Iroquois history, women have been intimately linked to corn, beginning with its arrival on Earth. They were entrusted with its care and became thoroughly intertwined with corn as sustainers of life.

In the Iroquois Creation Story, Sky Woman fell through a hole in the Sky World to the dark waters below, grabbing in her hands the seeds of several plants as she fell. In one version, a beast handed her a pot, a mortar and pestle, a marrowbone, and an ear of corn as she fell through the hole. Several geese caught her on their wings and gently deposited her on the back of a turtle where she began creating earth with a small handful of soil retrieved from the bottom of the ocean by a muskrat. By dancing, she expanded the area of the turtle's back and planted the seeds in the soil to initiate life on Turtle Island. In time Sky Woman gave birth to a daughter who later died giving birth to twin sons. From the daughter's grave grew corn, beans, squash, and sacred tobacco. Thus, corn emerged from the body of a woman, and its arrival in Iroquoia was intimately connected to the beginnings of life on Earth.

Corn appears again in the epic text of the Great Law, which describes the establishment of the Iroquois Confederacy. Iroquoia was in a state of almost constant warfare with communities taking up arms against each other in round after round of fighting and

> *"[The Thanksgiving Address] explains that people exist in reciprocal relationships with everything in the cosmos, each party having received their original instructions from the Creator."*

retaliation. The Peacemaker, a Huron from the north side of Lake Ontario, arrived in Iroquoia to propose a new governance system based on the ritual of condolence and community representation and participation in a central council. In the time of chaos and bloodshed, women had enabled the warriors to make war by providing them with food. The Peacemaker met with Jigonsaseh, a powerful corn grower from the West, and asked that the women stop feeding the warriors. Jigonsaseh agreed but negotiated an influential role for women in the Confederacy: clan mothers would appoint and depose the chiefs who represented the clans and nations in the central council. Therefore, women and corn played key roles as the Confederacy was established.

The Thanksgiving Address provides further evidence for the importance of corn. For hundreds of years, it has been recited in Iroquois communities at formal ceremonies, governmental functions, and other gatherings. Its use continues today in much the same way: to remind people of the essentials of life. The Thanksgiving Address outlines Iroquois worldview, describing the relationships between people and all parts of the natural world. It explains that people exist in reciprocal relationships with everything in the cosmos, each party having received their original instructions from the Creator. All parts of the universe are related as members of an extended and caring family. People and the crop plants corn, bean, and squash (called "Our Sustainers" or *Tsyunhehkwa*) share a unique relationship. These plants provide food for the Iroquois, who in turn give thanks, respect, and care to the corn, bean, and squash. Corn is further recognized as a plant that provides medicine for healing when people are sick. *Tsyunhehkwa,* also referred to as "the Three Sisters," are viewed as female members of the extended human family (see figure 2, page 6).

Figure 2 *The Three Sisters*, painting by Ernest Smith. *Published with permission from the collections of the Rochester Museum & Science Center, Rochester, NY.*

Corn Through Western Eyes

According to ethnobotanists, corn (maize) was originally developed by indigenous farmers in Central America and was transported over several millennia both north and south from its place of origin. Teosinte, a grass found in the central highlands of Mexico and Guatemala, was a wild ancestor of corn and was used in the domestication and breeding of corn by indigenous peoples from that region almost six thousand years ago. The transformation of teosinte into corn resulted in a tremendous increase in grain size and number, leading to enormous gains in food production for farmers who adopted it. As a result, corn was highly valued and moved rapidly across the major trade routes dissecting the Americas. It moved south into Peru, where it became a staple of the Inca Empire, and north through Mexico and into the Southwest and the Mississippi Valley of what would become the United States. In each of these areas, complex civilizations developed that depended on the cultivation of corn.

Archeologists now date corn's first appearance in New York to 100 BCE where it became an increasingly important source of food for Native communities over several hundred years. In other areas of the Eastern Woodlands, corn replaced other domesticated plants used by the region's earliest farmers, such as *Chenopodium berlandieri* (a close relative of common lambsquarters or goosefoot), *Iva annua* (sumpweed), and *Helianthis annus* (sunflower). But no evidence of these earlier domesticates has been found in New York. By the middle of the fourteenth century, a sophisticated polycultural cropping system that used corn, beans, and squash provided the economic foundation for Iroquois communities (see figure 3). These three crop plants were widely cultivated across the Northeast by the time the Confederacy was established, as described in the Great Law.

Figure 3 *The Three Sisters*, drawing by John Kahionhes Fadden (1988). *Published with permission from the New York State Museum, Albany, NY.*

Corn at Contact

By the time Europeans arrived in North America, corn was the staple food crop, cultivated in extensive holdings from Québec to Florida. In the Northeast during the sixteenth through eighteenth centuries, as European explorers and military conquerors moved inland from the New England coast, they described in their journal entries the widespread production of corn by Native peoples. Much of what would become New York state's prime farmland was first used by Iroquois farmers to grow vast acreages of corn and other crops.

Reports by Jacques Cartier, Henry Hudson, and the French general, Marquis de Denonville, described the vast and productive cropland that was planted, cared for, and controlled by Native women. Denonville claimed that during his military strike against the Senecas at Ganondagan in 1687, his troops destroyed more than 1.2 million bushels of stored grain corn in a five-day period. Colonial travelers in the seventeenth century reported getting lost in the enormous cornfields outside the Onondaga villages. Mary Jemison, a European woman who was captured by the Shawnee in 1758 and later adopted by the Seneca, described in her biography how Seneca women planted, tended, and harvested large acreages of corn in the decades preceding the Revolutionary War.

During the Revolutionary War, soldiers in General John Sullivan's Expedition provided detailed descriptions of Iroquois agriculture in their journals and diaries. As they pillaged and razed Indian villages in central and western New York, soldiers reported an extensive and extraordinarily productive agriculture throughout the Finger Lakes and west through some of the region's finest agricultural land. In addition to vast acreages of corn, soldiers described large orchards with well-tended peach and apple trees and a cornucopia of productive, diverse vegetable crops. Some of these crops, such as apples and peaches, were European in origin and had been incorporated into Iroquois agricultural systems (see sidebar 1).

After the War

The Revolutionary War not only displaced thousands of Iroquois people from their traditional homelands, but it disrupted and transformed Iroquois political, social, cultural, and

Sidebar 1

Descriptions of Iroquois Agriculture from the Sullivan Campaign, 1779

Maj. John Burrowes, August 30, 1779, Middletown:
"The land exceeds any I have ever seen. Some corn stalks measured eighteen feet, and a cob one foot and a half long. Beans, cucumbers, watermelons, muskmelons, cimblens are in great plenty."

Lt. Samuel Shute, Tuesday, September 7, 1779, Canadaasago:
"We found 200 acres of exceedingly good corn intermixed with beans and squashes pompions & a few potatoes."

Col. Daniel Brodhead in a letter to Gen. George Washington dated September 16, 1779, describing military operations in August, 1779:
"But immediately after ascending a high hill we discovered the Allegheny River & a number of Corn Fields, ... The Troops remained on the ground three whole days destroying the Towns and Corn Fields. I never saw finer Corn altho' it was planted much thicker than is common with our Farmers. The quantity of Corn and other vegetables destroyed at the several Towns, ... must certainly exceed five hundred acres which is the lowest estimate..."

economic institutions. None was transformed more dramatically than agriculture. State and federal governments, working in tandem with Christian churches, urged and coerced Iroquois women to relinquish their farming responsibilities to the men, in keeping with European social and cultural patterns. American politicians and religious leaders feared that Iroquois men, with greatly diminished land holdings to support their hunting and fishing, and the loss of status and prestige for men's activities that was a consequence of the Revolutionary War, might resist these changes militarily. They pushed Iroquois men to engage in what was the traditional source of social and economic power for Euro-American men: farming. The Seneca prophet, Handsome Lake, received visions in 1799, and as a result he too urged Iroquois men to take up farming as a survival response to the turmoil and dysfunction that resulted from the Revolutionary War.

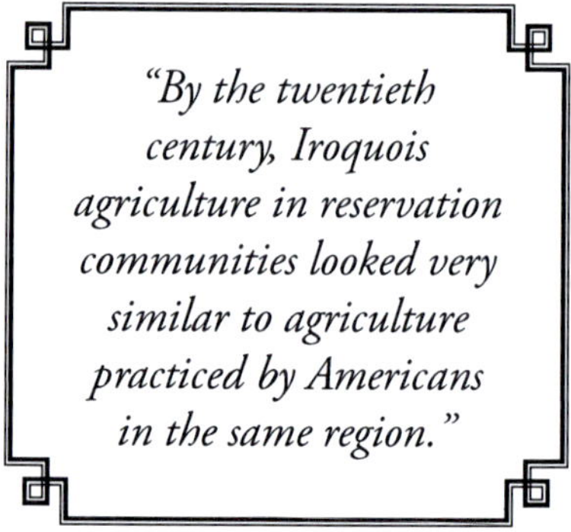

Although many Native peoples, both men and women, resisted these fundamental changes to Iroquois lifeways, Native communities in New York and Canada gradually transformed their agricultural practices. By the twentieth century, Iroquois agriculture in reservation communities looked very similar to agriculture practiced by Americans in the same region. Increasingly, Iroquois farms resembled the single-family households of their non-Native neighbors: men grew corn, wheat, potatoes, and other crops using plows and draft animals, while women raised children, tended to kitchen gardens, produced poultry, and engaged in spinning, weaving, basketry, and other domestic endeavors. Over generations, traditional family units and living arrangements, composed of a clan mother, her daughters, and their families, gradually disappeared and were replaced by households with a single nuclear family.

Corn and Its Influence on New York Agriculture

The changes affecting Indian agriculture in the area known as New York were not all one sided. From their first days in North America, Europeans ate Indian crops and adopted Native agricultural practices. Since corn was indigenous to the Western Hemisphere, it was unknown to the first colonists. But like millions of Native farmers before them, when introduced to corn, Euro-American farmers also found it enormously attractive and quickly added it to their agricultural repertoire. They rapidly learned to cultivate it successfully, and by the eighteenth century, most colonial farmsteads grew a substantial acreage of corn for both human and animal consumption.

Throughout the nineteenth century, American farmers continued to adapt and change traditional Indian corn varieties to meet the needs of their communities. Similar to Native farmers before them, they too produced a plethora of corn varieties adapted to particular microclimates, taste preferences, and uses. In New York state alone, scores of varieties were in active use by the end of the nineteenth century. During this same time period, government and university officials gradually developed a science-based agricultural research system in which improving the yield and agronomic performance of corn varieties was a primary goal.

By the end of the nineteenth century, American farmers regarded corn as a uniquely "American" crop, largely ignorant of its origins and the role that Native American farmers had played in its transformation to an American staple. Still, scientists in this time period referred to all corn varieties as "Indian corn." By the twentieth century, people began to use the term "Indian corn" to refer only to the multicolored corn varieties used to decorate at Halloween and Thanksgiving, unaware that all corn is "Indian corn."

By the late 1800s, most Americans were convinced that the traditional ways of the Iroquois, as well as all other Indian peoples across North America, would soon disappear. They expected that Indian peoples would assimilate into the larger Euro-American population and their languages, knowledge systems, and ways of life would no longer exist. Arthur Parker, a Seneca ethnologist who worked for the New York State Museum, began to record some of the most important aspects of Iroquois culture before it completely disappeared. In 1910 he wrote a monograph titled *Iroquois Uses of Maize and Other Food Plants* that detailed the practices the Iroquois had traditionally used to grow their crops. Because he had access to people living in many of the traditional Iroquois communities in New York and Canada, he was able to compile a thorough description of traditional Iroquois agriculture. Although Iroquois people at this time still grew and used traditional varieties of corn for use in ceremonies and to make soup, bread, and other traditional foods, they increasingly relied on American varieties for planting large fields of "production" corn.

Parker listed thirteen varieties of corn used by the Iroquois. A little more than a decade later, F. W. Waugh produced another monograph for the Canada Department of Mines that also provided descriptions of Iroquois corn varieties with a color plate showing several varieties (see figure 4). A comparison of their varieties is provided in Table 1.

Figure 4 **Some Iroquois Corn Varieties.** *Photo © Canadian Museum of Civilization, S86-1212. Published with permission.*

Table 1 Iroquois corn varieties as categorized by Parker and Waugh.

Parker	Waugh	Parker	Waugh
Flour corns		Flint corns	
Tuscarora	Tuscarora short	Hominy	White short
Tuscarora short ear	Tuscarora long	Hominy long ear	White long
Purple	Purple bread	Calico	Variegated
Red	Red bread	Yellow	Yellow
—	Variegated calico	—	Purple short
Popcorns		Pod corns	
Red	Red rice	Sacred, original	Sacred, original
White	White	Sweet corns	
—	Red pearl	Sweet	White, short
		Black sweet	—

Iroquois Corn in the Twentieth Century

Throughout the first half of the twentieth century, Iroquois communities continued to adopt Euro-American farming practices. Like their non-Native neighbors, they grew small grains such as wheat, oats, barley, and rye, along with corn, potatoes, and a diversity of other fruit and vegetable crops. Many farmers added livestock (primarily dairy) in this period as well. Tuscarora Throughout the first half of the twentieth century, Iroquois communities continued to adopt Euro-American farming practices. Like their non-Native neighbors, they grew small grains such as wheat, oats, barley, and rye, along with corn, potatoes, and a diversity of other fruit and vegetable crops. Many farmers added livestock (primarily dairy) in this period as well. Tuscarora farmers became noted for their fruit trees and Mohawk farmers for their dairy farms.

Erl Bates, who directed a Cornell Cooperative Extension program for Iroquois communities from 1920 through 1964, urged Indian farmers in the 1930s to stop planting their traditional va-

> *"[A]gricultural societies...encouraged Iroquois farmers to participate in agricultural fairs and supported agriculture as a means of improving reservation life."*

rieties in favor of the more productive ones that American farmers used. Although Indian farmers grew these improved varieties, many families continued to grow small acreages of traditional corn varieties for use in traditional foods and ceremonies. Bates was also instrumental in helping Iroquois communities form agricultural societies, which encouraged Iroquois farmers to participate in agricultural fairs and supported agriculture as a means of improving reservation life. Under Bates' leadership, the Six Nations Agricultural Society was formed, and it organized the Indian Village at the New York State Fair, which continues to this day.

Beginning in the 1930s, the number of farms and people engaged in agriculture steadily declined throughout the Northeast. Every year fewer people chose farming as a livelihood, and many established farms were lost because of retirement or bankruptcy. In Iroquois communities, agriculture similarly declined. Increasingly, Iroquois young people, like their non-Native counterparts, chose other careers, abandoning agriculture. Cornell University's agricultural extension activities in Iroquois communities became less and less frequent and eventually stopped completely by the time of Bates' retirement in 1965. A survey by the American Indian Agriculture Project at Cornell in 1989 found less than ten full-time farmers among all the major Iroquois communities in New York and Canada.

But despite the persistent decline in agriculture in Iroquois communities, even in the 1950s and 1960s individuals and families continued to grow traditional varieties of corn in home gardens and on small acreages, supplying their communities with corn for ceremonies and traditional corn-based foods. Beginning in the 1970s, Iroquois communities began to reassert their political and cultural sovereignty, reclaiming many of the traditions and institutions that had been damaged and marginalized by centuries of colonization. As these communities pushed to control their own school, healthcare, political, judicial, and economic systems, there was also resurgence in preserving and learning traditional languages and revitalization of traditional agricultural crops and foods.

Today, many Iroquois communities have several individuals or families producing traditional corn varieties (and other crops as well). In some cases, the growing of traditional crops is an organized community or tribal government activity. Although many of the traditional corn varieties have been lost or are severely endangered because there is so little seed available, Tuscarora (or Iroquois) white flour corn is still widely available in Iroquois communities.

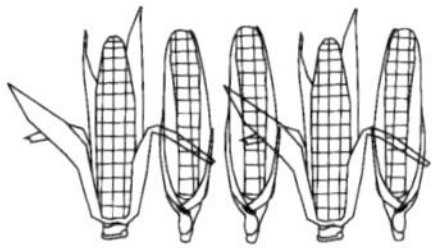

Botany of Corn

> *"The vast majority of corn grown in the United States today is hybrid dent corn... People all over the world prize flint, flour, and dent corns, in a range of colors..."*

Corn Types

Agronomists recognize five main types of corn, based on the characteristics of the endosperm (starch) within the kernel. The color of the corn kernel is relatively unimportant in identifying corn varieties or types. Corn endosperm can range from soft and floury to very hard or corneous. *Flour* corns have predominantly soft endosperm, and when ground they produce smooth flour. In contrast, *flint* corns have hard endosperm, while *dent* corns exhibit a mixture of soft and hard endosperm. Among these three types of corn, the hardness of the endosperm varies across a continuum, so that sharp delineations between them can be difficult. *Popcorns* have an extremely hard, corneous endosperm that traps water until it is heated and the erupting steam breaks open the endosperm. *Sweet* corns, with a shrunken endosperm, are composed primarily of sugars rather than starch (see figure 5).

The vast majority of corn grown in the United States today is hybrid dent corn, of which approximately 75 percent is used to feed animals. Still, industrial and food products represent an important use of corn. In the United States, tortillas, corn chips, cereal products, sweeteners, and alcohol now comprise an increasing portion of corn production. And in much of the developing world, corn serves as a major cereal grain for human consumption. People all over the world prize flint, flour, and dent corns, in a range of colors, for their food value, as well as for their unique flavors, textures, and cooking characteristics.

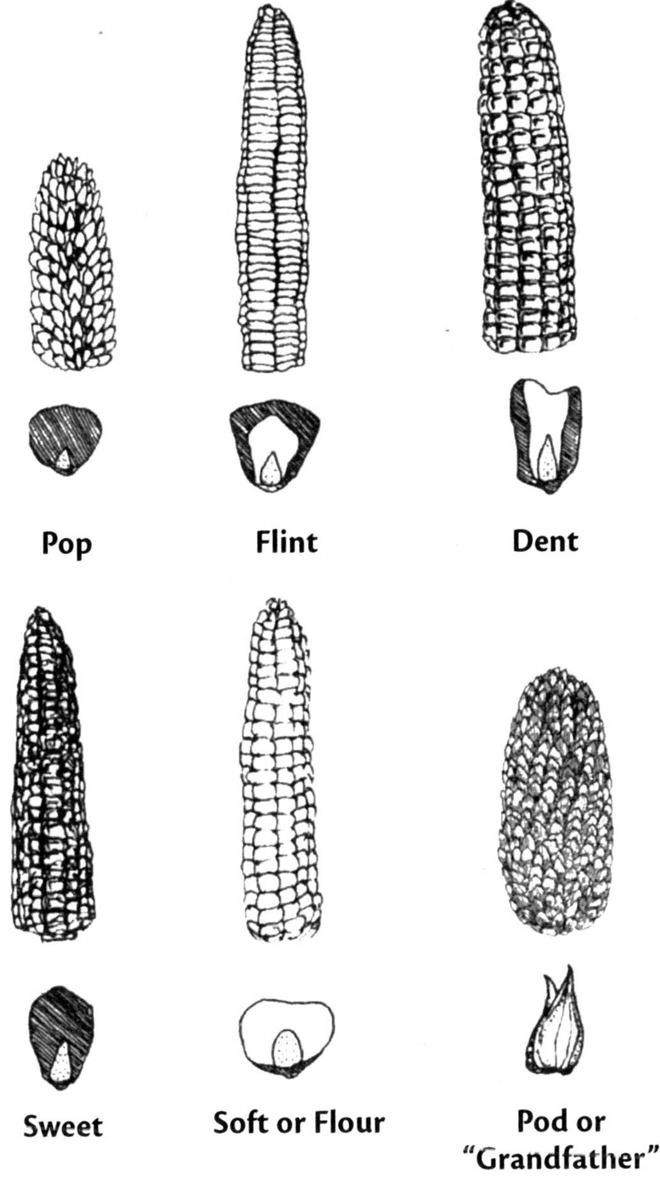

Figure 5

Corn Types or Races. Credit: Marcia Eames-Sheavly. *The Three Sisters: Exploring an Iroquois Garden.* Ithaca, NY: Cornell University Cooperative Extension, 1993. *Reprinted with permission.*

Open-Pollinated vs. Hybrid Corn

Until the twentieth century, farmers and corn breeders relied on the corn plants' natural process of cross-pollination to select new characteristics and to improve corn yield and other qualities. The corn plant, with separate male and female flowers on the same plant, relies on windblown pollen from the tassels (male flower) of one plant to fertilize the individual silks (female flowers) of any corn plant in the field. When this process is allowed to occur naturally (i.e., any pollen from any plant is able to fertilize any silk in the field without obstruction), this corn is referred to as open-pollinated corn. Seed from the most desirable plants and ears is saved for next year's planting.

In the 1920s, agricultural scientists discovered that significant yield increases could be obtained by first producing inbred lines of corn in which the pollen from one plant is used to fertilize its own silk. Scientists used this process of inbreeding for several generations to obtain plants with desirable and stable characteristics. When two inbreds were crossed, they sometimes produced a hybrid with higher yields than either of the inbred lines (see figure 6). In addition, within a specific hybrid variety, all plants were genetically identical so they were very uniform, in contrast to an open-pollinated variety whose plants could exhibit significant variability. To produce both the inbred line and the hybrid, pollination was controlled: only pollen from particular plants was allowed to fertilize the silks. Farmers, however, were advised not to save or plant the seed from these crosses because yields usually declined in the next generation by 20 percent or more. As a result of this research on hybrid corn, a hybrid seed industry has emerged that produces and sells new seed to farmers each year.

Continued breeding efforts with hybrids from the 1920s through today have produced high-yielding varieties with increased stalk strength and resistance to various diseases and pests. These characteristics, coupled with the inherent uniformity of the hybrid corn plants that facilitates mechanized harvest, have resulted in large increases in U.S. corn production. Research on open-pollinated varieties essentially ended in the 1940s since hybrids had begun dominating U.S. cornfields. Some people believe that had research on these varieties continued with the same intensity given to hybrids, open-pollinated varieties would now be much more competitive in yield and other agronomic characteristics, compared to hybrid corn.

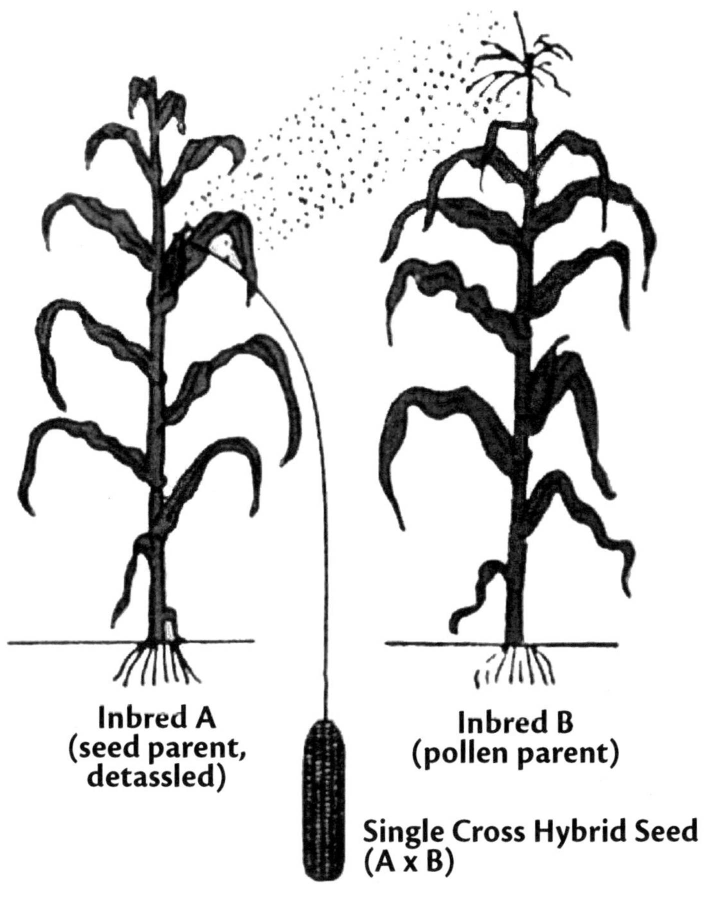

Figure 6 **Modern Hybrid Corn (Single-Cross Hybrid Seed).** Modern corn hybrids are produced in a two-step process. First, two parent plants are selected which are repeatedly self-pollinated (pollen from the corn plant is used to fertilize the silk of the same plant) until no new characteristics appear in the offspring of the self-pollinated plant. Then, pollen from each of the self-pollinated plants (Inbred B) is used to fertilize the silks of the other plant (Inbred A). The resulting hybrid (Single-Cross Hybrid Seed [A x B]) will combine characteristics of each parent and will have higher yields than either of the original parent plants.

Credit: C. Wayne Smith, Javier Betran, and E.C.A. Runge (eds.). *Corn: Origin, History, Technology, and Production*, Wiley Series in Crop Science. Hoboken, NJ: John Wiley & Sons, Inc., 2004. *Reprinted with permission of Wiley-Blackwell, Ltd. a subsidiary of John Wiley & Sons, Inc.*

Today, virtually all corn acreage in the United States is planted with hybrids that are the result of controlled crosses of inbred lines. However, farmers in developing countries still grow open-pollinated corn because it is the least expensive way to produce corn, and it allows them to improve agronomic characteristics in the absence of a developed seed industry.

Maintaining "Pure" Varieties

Many growers today are concerned about contaminating open-pollinated corn through inadvertent cross-pollination. Prior to the introduction of hybrid corn, every corn plant in a field was genetically different from its neighbor. Although every kernel on an ear has the same female parent, it is possible for each kernel to have been pollinated by a different male parent. The resulting variability of this "open pollination" provides an enormous pool of genetic material that has been used by farmers to develop varieties that differ dramatically in their characteristics, including kernel type and color, plant architecture, and environmental response. Individual farmers, using this huge pool of genetic variability within each corn plant, have developed varieties of corn with characteristics adapted to an enormous array of climates and specific uses. As a result, today there are literally thousands of

Figure 7 **Calico Corn**. *Courtesy of Jane Mt.Pleasant.*

Figure 8 **Blue Corn.** *Courtesy of Jane Mt.Pleasant.*

corn varieties adapted to and grown in different environments, from lowland tropical forests to high-altitude plains, and with uses ranging from sweet corn, to tortillas, to popcorn, to corn feed for hogs, poultry, and cattle. Corn can grow on stalks as short as thirty inches to as tall as almost twenty feet, with kernel colors that are white, black, purple, red, orange, blue, calico, maroon, and many other colors (see figures 7 and 8). Many additional differences are invisible, meaning that they are not expressed in the plant's appearance, but the genes for characteristics are still present in the plant's genetic material and can be passed on to its offspring through sexual reproduction.

What Is a Variety?

Since every plant in an open-pollinated cornfield is genetically different, it is impossible to have a "pure" corn variety: all the plants differ from each other in some way. In a field of open-pollinated corn, the majority of ears and kernels will be similar to what the farmer or gardener planted, but many of the plants will show small differences or variation in stalk size, ear placement, kernel characteristics, etc. These ears and kernels will still be recognizable as fairly similar to the variety that was planted, but they are not the best representatives or most characteristic of that variety.

A few of the kernels and ears in an open-pollinated cornfield may be strikingly different from what was planted. For example, in a field of blue flour corn, you may find individual white sweet corn kernels on some ears. This doesn't necessarily mean the corn was contaminated by cross-pollination with white sweet corn, although that is one possible explanation. Instead, it may indicate that within the genes of one of the thousands of possible parents in that cornfield, there were some sweet corn characteristics and some white colors that were invisible in the ears/kernels of the parents. However, they appeared in a very small number of offspring kernels. In these cases, the white color and sweet endosperm that had been invisible in each of the parents was expressed during cell division in a few of the hundreds of thousands of individual kernels that developed in the field. It is much like the case of two brown-eyed parents who have a child with blue eyes. Although other explanations are possible, it may simply indicate that both parents had genes that allowed the blue color to appear only when these genes were combined with each other in the child. Similarly, only a few combinations of genes from the two corn parents enable both sweet endosperm and white color to be expressed. In most of the possible combinations of genetic material from the two parents, the endosperm will be floury and the kernel color blue.

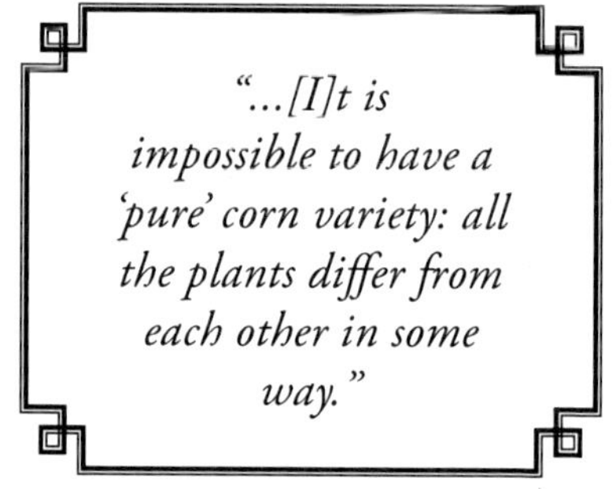

> *"Since there is an enormous amount of variability within the germplasm of all the individual corn plants, the characteristics of the variety can change over time."*

Farmers as Corn Breeders

Growers who save their own seed usually have an idea of what a particular variety should look like, and they choose plants, ears, and kernels for next year's seed that most closely match their image of the ideal form of that variety. But unless all farmers growing one variety of corn jointly decide on the specific characteristics they will choose in selecting seed, after several years, their corn crops may look quite different. Since there is an enormous amount of variability within the germplasm of all the individual corn plants, the characteristics of the variety can change over time. Thus, farmers who start with corn seed from the same variety can, over the course of several years, end up with varieties that are distinctly different.

This is probably not the result of contamination. It simply reflects the enormous genetic variability within corn plants and the ability of individual growers to change the plants' characteristics through their selection of seeds. In this way, growers have maintained characteristics they like and associate with a particular variety (i.e., red cobs at least twelve inches in length) and also introduced new characteristics (i.e., early maturation or multiple tillers). It is this combination of variability within the genetic makeup of corn and the responsiveness of corn to human intervention (i.e., selection of seed) that has made corn so attractive to farmers around the globe. Farmers use seed selection to tap the enormous pool of characteristics within the corn plants' genetic material to develop varieties for an enormous range of environments and needs. They can then maintain these varieties over time by carefully selecting plants, ears, and kernels that have those characteristics. Every grower becomes a plant breeder when he or she selects certain ears to save for seed corn for the following year's crop.

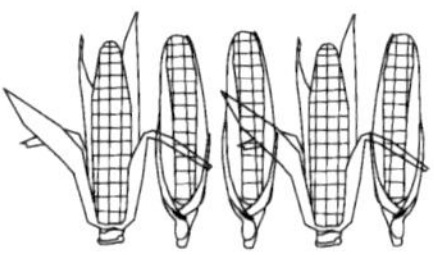

Growing Corn

Growing Iroquois corn is similar in most ways to growing commercial hybrid corn. Growers need to carefully select fields, supply nutrients in the appropriate amounts and at the appropriate time, protect the corn against pests, such as weeds, insects, and animals, and plan for the harvest and storage of the grain.

Differences Between Iroquois and Hybrid Corn

However, Iroquois corn also differs significantly from hybrid corn in several respects. These differences include 1) greater need to protect corn from cross-pollination by unwanted varieties (hybrids or other open-pollinated corn), 2) lower recommended plant populations, 3) less nutrient requirements (N, P, and K) because of lower yields, 4) poorer standability and disease resistance, 5) different harvest and storage techniques required for seed saving, and 6) more difficult mechanized harvest due to variability in height, ear placement, number of ears, kernel size, and maturity. Some of the specific concerns and issues for Iroquois corn production are described in detail below. Since people interested in growing traditional Iroquois corn may be planting in home gardens or on a larger scale, information is provided for both situations.

For general information on managing soils and crops in your area, consult your local Cooperative Extension office. For example, in New York, the *Cornell Field Crops and Soils Handbook* provides detailed information on the state's soils and crops. Gardeners and growers can also contact the Cooperative Extension office in their counties for specific advice and recommendations on growing open-pollinated corn in their area. Other sources of information are listed in the References section.

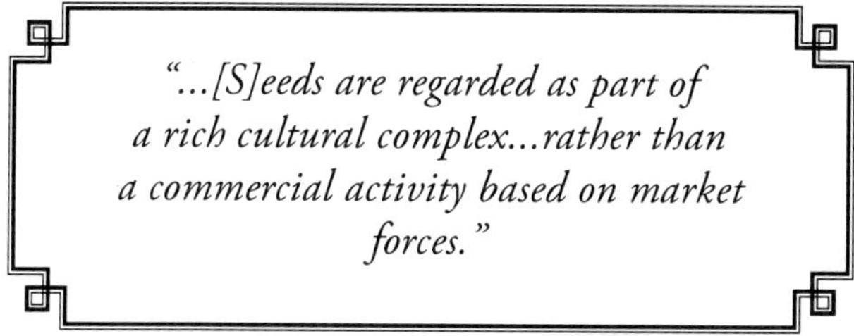

Finding Seed

Seed from traditional Iroquois varieties can be most easily obtained from individuals in Iroquois communities who are currently growing traditional corn. A few seed companies and other organizations also sell one or two of the more widely available varieties, particularly Tuscarora White Flour (also called Iroquois White Flour).

Many Native peoples today are unwilling to market the seeds of traditional crops and other plants commercially. This reluctance springs from two sources. First, within Iroquois communities, seeds are regarded as part of a rich cultural complex and their exchange is part of the fabric of the community, rather than a commercial activity based on market forces. Second, indigenous communities have been exploited repeatedly, with others profiting from Native land, plants, objects, and knowledge. Consequently, many communities are increasingly reluctant to give or sell seed to people they don't know.

What if you want to grow these crops, but don't live in an Iroquois community, or don't know someone who currently grows traditional Iroquois corn? Many seed companies and individual farmers sell open-pollinated corn varieties that are similar to traditional Iroquois varieties. If you select varieties that are suited for your climate and growing season, they will perform much like the traditional Iroquois corn. (If you are a gardener or farmer outside of the Northeast, the Iroquois varieties would probably not perform well in your area anyway.) Using these commercially available varieties, you can, as described above, develop your own open-pollinated corn suited for your microenvironment and particular needs. These corn varieties can also be processed for human food, in the same ways described below. In fact, all of the information about growing traditional Iroquois corn in this book is applicable to all open-pollinated varieties. A growing community of people across the United States and Canada are producing open-pollinated corn. You can find information in the References section on how to connect with them.

Field Selection

Your first task as a corn grower is to find the right place to plant. Corn grows best on level or gently sloping fields with moderate to good drainage and a soil pH of 5.8 or higher. For home gardens, select sites that receive at least six hours of full sun every day and that do not have wet or boggy areas. In order to minimize rootworm problems, do not plant corn following corn. Plant another crop in the field for at least one year before you plant corn there again. If you want to avoid mixing with other corn varieties, you should plant corn on a site that is isolated from fields of hybrid corn or other open-pollinated corn varieties. In general, corn should be planted at least 600 to 1,000 feet from other cornfields to prevent pollination with those plants. If you are going to save your own seed, you may want to increase that distance. More information on this is provided below.

Fertility Management

Soil samples should be taken from the field or garden and analyzed prior to planting. Your Cooperative Extension office can provide instructions on how to take and submit soil samples and can assist in interpreting the results. Liming and fertilization recommendations can then be developed, depending on whether you will be using organic, chemical, or a combination of these fertilization practices. Fertility recommendations will usually be lower for Iroquois corn than for hybrid field corn grown on the same soil because of the lower yield potential of Iroquois and other open-pollinated varieties. However, they may have similar fertility recommendations as sweet corn.

You can use green manures, cover crops, animal manures, compost, organic amendments such as bloodmeal, and/or inorganic fertilizers to provide nutrients for the corn. Many farmers supply the corn's nitrogen requirement by side-dressing before the last hoeing or cultivation. Cover crops are particularly beneficial because they build soil organic matter, add essential plant nutrients, reduce soil erosion, and break pest cycles. Red clover, hairy vetch, rye grain, and ryegrass are all excellent cover crops that are well adapted to the range of conditions in the Northeast, but only the legumes (clovers and vetches) will provide nitrogen for the corn. (For areas outside the Northeast, you should contact your Cooperative Extension office for recommendations.) One year's growth of red clover, when turned under, usually releases sufficient nitrogen for a corn crop. If you are not using compost, animal manure, or inorganic nitrogen fertilizer, you should use a leguminous cover crop to provide sufficient nitrogen to the corn. One year of a cover crop (or any crop other than corn) will also break the life cycle of corn rootworm in the Northeast. See the References section for information on using cover crops.

Cropping System/Rotation

Decide how the corn will fit with other crops that you grow. Diverse cropping rotations are needed in both home gardens and larger operations. If you grow corn organically, a crop rotation, including legume/grass hay and/or cover and green manure crops, is essential, particularly if you don't use manure or compost. There are many publications that address soil and crop management using organic methods (see the References section). Even if you intend to use inorganic fertilizers, developing a diverse crop rotation, rather than continuous corn, will reduce insect, disease, and weed problems in both gardens and larger fields.

Working the Ground

For field-scale operations, two forms of tillage, primary and secondary, are normally performed before planting. The goal of tillage is to prepare a reasonably smooth and friable seedbed so that the corn can be accurately planted and covered, and to remove any weedy vegetation that will compete with the corn seedlings. Primary tillage, either moldboard or chisel plowing, inverts and mixes the soil, as well as incorporates residue from the previous crop and uproots existing weeds. Secondary tillage, which includes disking and harrowing, breaks up clods and smoothes the soil surface.

Reduced Tillage and No-Till

Although conventional tillage practices (i.e., moldboard plowing, etc.) provide an excellent seedbed and bury weedy vegetation, they also create some conditions for excessive soil erosion. As a result, many farmers are reducing the intensity and number of their tillage operations. *Reduced tillage* methods often include chisel plowing followed by a single pass with a secondary tillage implement or a single pass with a combination implement. *No-till* systems require a planter capable of tilling a narrow strip of soil and then placing the seed in land that is left unplowed except for the strip of soil moved by the planter. The larger amounts of plant residue left on the soil surface with reduced or no-till systems offers more protection from soil erosion than with conventional tillage. However, as you reduce tillage, weed control becomes more difficult. With reduced tillage, you may need a special high-residue cultivator to control weeds mechanically. No-till systems often rely heavily on herbicides for weed control. Additional resources for reduced tillage farming are listed in the References section.

> *"In home gardens...[t]he garden can be rototilled to provide easily worked soil...Corn can then be planted in rows or in mounds..."*

In home gardens, growers also have several options. The garden can be rototilled to provide easily worked soil. You can mix in compost, fertilizers, manures, and/or lime and simultaneously eliminate weedy vegetation in one operation. Small areas can be spaded manually, removing weeds as you go and mixing in fertilizers, compost, or manure. Corn can then be planted in rows or in mounds, as described below.

If you want to reduce the amount of soil disturbance in your home garden, planting in mounds is the best option. The first time you form mounds, you will need to either rototill or spade up the soil, but in subsequent years, the area between the mounds can be left undisturbed and only the soil directly within the mound is moved during planting. Detailed instructions for forming mounds are given below. But note that mound planting encourages continuous corn, which may not be desirable.

Planting: How, When, and How Much?

In the Northeast, corn should be planted in early- to mid-May to take full advantage of the growing season in this area. In large fields, plant at a lower population than you would hybrid field corn. Start with 16,000 to 20,000 plants per acre until you gain experience with the variety. For home gardens, you can plant the corn in rows or mounds. If planting in rows, allow thirty-six inches between rows, and plant one kernel every eight to ten inches within the row.

If you make mounds, space them approximately three feet from mound center to mound center. Initial height of the mound should be four to six inches above the soil surface with a diameter of eighteen to twenty-four inches. Plant four to six kernels per mound, and when the seedlings emerge, decrease to three plants. If you use mounds over several years, they will grow each year as you add plant residues and hoe up the soil into the hill during planting and weeding. This added vegetation not only provides soil nutrients as it decomposes, but over time it will also improve the tilth and friability of the soil as organic matter levels within the hill increase. The resulting improved drainage, aeration, and water-holding capacity all favor better corn growth.

Planting Equipment

Small plantings of corn may be made by hand with hand jabbers or with push-type seeders. Large plantings generally require a mechanical corn planter, although fields as small as an acre or two can be planted with hand jabbers (see figure 9). For mechanized planting, if the seed has been graded for size, you can use seed plate, air, or vacuum planters. These planter types require seed that has been graded for size in order to match the seed plate or disc to the size of the

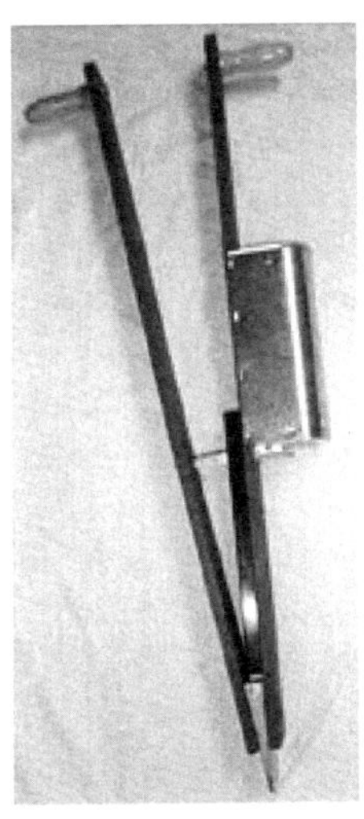

Figure 9 **Corn Hand Jabber.** *Courtesy of Jane Mt.Pleasant.*

corn seed being planted. Corn seed that is ungraded will be a mixture of several sizes of flat and round kernels. Finger pickup planters usually perform well with ungraded seed. Air or vacuum planters may also perform well. With plate-type planters, selecting the proper seed plate for ungraded seed may be troublesome. Seed cells that are too small will cause skips or may break kernels, while cells that are too large will drop doubles. Doubles are usually preferred to skips.

Corn planted early (before mid-May in New York) should be placed one to one-and-a-half inches deep in the soil. Later plantings made when soils are warmer and often drier can be two to two-and-a-half inches deep. If you use inorganic fertilizer, band it two inches to the side and two inches below the seed when using a corn planter, or mix it in well with the soil when planting by hand.

Weed Control

How well you control weeds may be the single most important factor in the success of your corn crop. In fact, it should be the first thing you think about when selecting the site. The type and quantity of vegetation that is already growing in the field or garden will give you a clue as to what you can expect after the corn is planted. Perennial weeds such as quackgrass and nutsedge can rob water and nutrients from young corn plants, preventing them from growing and developing normally. If you have a field or garden full of quackgrass, plan to suppress it or kill it before you start planting corn. Otherwise, you face a season-long battle with this weed that you will likely lose.

Annual weeds can also completely overtake a field or garden, shading out the young corn plants that may not be quite so vigorous or numerous. Huge flushes of annual weeds like common lambsquarters, foxtails, and pigweeds can come in waves throughout the first six weeks after planting. If you don't control them, your corn plants may yield little or nothing. So inspect the garden or field site well before planting and put together a weed management plan before you plant the first kernel of corn.

For home gardens, hand weeding is probably the best option, but if you have heavy infestations of perennial weeds like quackgrass, you may want to lay down heavy plastic or other types of mulching material the fall before to at least weaken and perhaps kill the quackgrass before you do any tillage in the spring.

Although moldboard plowing and even chiseling will remove the surface growth of quackgrass, these perennial plants quickly reestablish from underground rhizomes that are not damaged by plowing. If the rhizomes are not killed, new plants emerge from the rhizomes and quickly reinfest the area. Annual weeds in home gardens can usually be controlled by diligent hoeing and/or the use of organic mulches.

For field-scale operations, the options are more complicated. Both chemical and mechanical weed control strategies can be used, depending on your preferences. Unlike modern hybrids, open-pollinated corn varieties emerge irregularly and vary considerably in height, factors that can affect herbicide applications and mechanical cultivation. Many sources for weed control information are given in the References section. Regardless of your control method, weeds should be removed early in the growing season to prevent competition and to allow the corn to grow ahead of the weeds.

Harvest

Except for sweet corn varieties, all Iroquois corn should be harvested as hard, dry kernels that have less than 25 percent moisture. Corn will mold easily in storage if it is too wet, so it is important to know the moisture content of the kernels when you harvest. Commercial growers use a grain moisture meter, but these are generally too expensive for home gardeners. If you live in an agricultural area, you may be able to take a few ears to a local farmer, feed store, or grain dealer where they can easily determine the moisture for you (see sidebar 2, page 32).

For most areas in the Northeast, harvest will take place in October, although it may extend well into November. Home gardeners can pick ears as they dry down to the safe range and then store the ears in a dry, cool area that is protected from insects and other pests, such as squirrels and mice. Growers with larger acreages can also pick and store husked ears.

Ears can be harvested by hand or with a mechanical picker that picks ears and removes husks. Traditionally, Iroquois farmers have picked the corn by hand and then braided (see figure 10, page 32.) the ears and hung them in a covered area protected from rodents and other pests. Ears can also be stored in mesh bags and hung in a ventilated building. The corn continues to dry down to

Sidebar 2

Estimating Corn Grain Moisture

First, examine the corn kernel for the black layer at the base of the dissected kernel, which indicates that it is physiologically mature. Pinch individual kernels as hard as you can. If they release any moisture (milk), they are likely above 35 percent moisture and should not be harvested. Kernels that are pliable or slightly soft as a result of hard finger pressure, even though they don't release milk, are probably between 25–30 percent moisture. You should allow the corn to dry more before harvesting. Corn grain is ready to harvest when you can see the black layer and the kernel is very hard and brittle.

Figure 10 **Corn Drying in Braids.** *Courtesy of Jane Mt.Pleasant.*

13–15 percent moisture as it hangs in storage.

Ears of corn can also be stored in cribs where they will gradually continue drying to 13–15 percent grain moisture (see figure 11). Grain moisture must be below 25 percent at harvest in order to store corn safely in cribs. With the Northeast's humid climate, a crib should be no more than four-and-a half feet wide. Because the ears complete drying during storage, the crib needs good exposure to the prevailing wind. Minimize husks, silks, fine particles, and other debris that impede air flow through the cribbed corn. It is also important to protect cribbed corn from rodent damage. Ear corn can also be artificially dried, but it is better to select varieties that will mature and dry in the field to moisture levels that are safe for crib storage. See the References section for information on building a corncrib.

Combining corn in the field is not recommended because of the difficulty caused by ear molds, described below. But corn ears can be shelled mechanically or by hand after they have dried to 12–15 percent moisture in cribs or on braids. Then they should be stored in airtight bins. Homebuilt granaries or barn floors can

Figure 11 **Daniel Winnie's Family, Six Nations Indian Reserve, ON.** Photo depicts an example of a corncrib. *Photo by F. W. Waugh, 1912, 17145, published with permission from the Canadian Museum of Civilization.*

also be used for storage. Dry corn weighs forty-three pounds per cubic foot at 13 percent moisture, so make sure that the floor of your storage building will support this weight.

Ear Molds

Many of the traditional varieties of Iroquois corn are susceptible to ear mold diseases (see figures 12 and 13). Flour corns are particularly vulnerable to these infestations. The most common molds are *Fusarium graminearium* (*Giberella zeae*) and *Fusarium moniliforme*. Some are highly toxic and make both people and animals very sick. Moldy grain should never be eaten or fed to animals. Inspect ears carefully as you braid or place the ears in a corncrib, discarding ears with moldy kernels. Inspect the ears again when you shell them for use or for further storage.

Figures 12 and 13 **Ear Mold Infesting Corn Cob.** *Courtesy of Robert F. Burt.*

Saving Seed Corn

Grain from open-pollinated varieties can be saved to plant the following year. All varieties of corn will readily cross, so make sure that fields from which you intend to save seed are isolated from other varieties, whether they are open-pollinated or hybrid corn. Without proper isolation, pollen blown in from adjacent fields may pollinate your seed corn. Other cornfields should be at least 600 feet away, but one-fourth of a mile is better. Staggered planting dates, use of woods and hedgerows, and consideration of prevailing winds will also help prevent unwanted cross-pollination. (Cross-fertilization is of little or no concern for corn that will be eaten rather than used for seed because the occasional kernel that results from unintended pollination has almost no effect on the taste or quality of the corn.)

Select ears to be saved for seed from plants in the field *before* you harvest the rest of the field. Although you can select ears after the field has been harvested, you lose the opportunity to identify the plants that have important traits like early flowering, strong stalks, and healthy leaves. Select plants when the husks have dried but leaves and stalks are still partly green. Choose plants in the center of the field that have a uniform stand of plants on all sides. Identify sturdy, well rooted, upright, and disease-free plants; from these plants, select ears with favorable characteristics. Take ears only from plants that are free from leaf diseases, insect damage, mold, and lodging.

Ears selected for seed should have characteristics that you consider desirable. These may include ear position, husk coverage, ear length, row number, and kernel size and color. Most likely you will select ears that conform to your idea of what this particular variety should look like. Remember that every grower becomes a plant breeder when he or she saves seed. You may, over time, end up with a variety that looks somewhat different from the "same" variety produced by another grower.

In order to maintain genetic diversity, don't select too tightly for a particular ear type; in other words, pick ears with a range of the characteristics you consider most desirable. In this way, you maintain a variety with greater diversity in its pool of germplasm. It will be less susceptible to insects, disease, and climatic fluctuations because of the diversity in its gene pool. In order to prevent inbreeding, select at least one hundred ears from which you will save seed, and pick these ears from several areas in the field. Over a number of years, careful selection can produce strains of a variety that are well adapted to your particular environment and location yet still retain the essential characteristics of that variety.

Ears selected for seed must be carefully dried to maintain germination and preserve quality. Don't shell the ears until they have completed air-drying. As described earlier, ears can be braided and hung to slowly air-dry in barns or other well-ventilated buildings. You can also simply put the ears in mesh bags (onion bags work well) and hang them in buildings or on top of cribbed non-seed corn. Small quantities can be dried right in the house. If large quantities of seed corn are harvested, you can also store them in separate cribs or sectioned cribs. If you dry the seed corn artificially, do not use temperatures above one hundred degrees Fahrenheit.

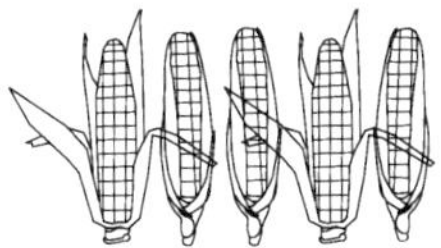

Eating Corn

Corn can be eaten many ways, providing an array of nutritious and appealing dishes, depending on the type of corn used and when it is harvested. Foods prepared from corn that has been allowed to fully mature, producing hard, dry kernels, provide the most food value in terms of high-quality nutrition and energy (calories). Outside the United States, whole-grain dishes prepared from flour, flint, or dent corns provide the essential calories and nutrients for much of the world's population.

Dishes prepared from corn harvested in the milk stage, when the kernels have not yet reached physiological maturity, are relished for their distinct flavor and sweetness. Sweet corn is almost always harvested in the milk stage when it is still immature and few of its sugars have changed into starch. Most people in the United States eat primarily sweet corn, which is lower in starch, and whose kernels consist mostly of sugar and water. Although the flavor is delightful, the food value of sweet corn is modest. There are several traditional varieties of sweet corn (see table 1, page 13), as well as traditional Iroquois popcorn varieties, which are eaten primarily as a snack food. However, Iroquois people are more likely to eat food prepared from flour and flint corns.

Flour and flint corns are usually harvested when mature, but several dishes made from flour or flint corn picked in the milk stage are treasured by Iroquois people for their delightful flavor. Corn picked at this milky stage must be processed immediately, in contrast to mature corn kernels, which when dried to 12–15 percent moisture, will keep for several years if protected from rodents and insects.

Flour and flint corns are usually eaten as whole kernels in soups or stews, or ground into flour and prepared as bread or mush-like cereals. Corn soup, a traditional staple in Iroquois diets, is prepared by boiling the corn first with wood ash to remove the hull or pericarp, and then after several rinses, cooking it with meat, beans, and other savory additions.

Nixtamalization or Alkaline Cooking

Cooking corn with wood ash (or other high pH substances such as lime, baking soda, or lye) is called "nixtamalization" in Mexico where it is thought to have originated. The process is also referred to as alkaline cooking. In Mexico, corn is boiled with a limestone solution and then ground into a dough called nixtamal, from which tortillas are prepared. All indigenous peoples in the Western Hemisphere who rely on corn as their major food source use alkaline cooking to prepare corn for consumption.

In Iroquoia, corn is boiled with wood ash, which provides several benefits. First, it reduces the cooking time of corn by removing the hull or pericarp, which protects the kernel from damage and enables the corn to be stored for long periods of time. Although a hard hull is advantageous when storing corn, it greatly increases the time required to cook the kernel to the point where it is soft enough to chew easily. Alkaline cooking chemically loosens the hull from the kernel, allowing the kernel to more readily absorb water and soften during cooking. In the United States, whole kernel corn that has been cooked in an alkaline solution to remove the hull is called hominy.

Second, and perhaps more importantly, cooking with wood ash improves the nutritional value of corn by increasing the availability of two essential nutrients in the human diet: amino acids and niacin. Nixtamalization improves the balance of amino acids, the building blocks of protein, making more of them available to humans. Since both the quantity and quality of protein in corn are somewhat marginal (see "Corn and Human Nutrition"), this is very important for people who rely on corn for most of the protein in their diet. Cooking corn in an alkaline solution also makes some of its bound niacin more available. Niacin is a component of coenzymes responsible for cellular oxidation and respiration. The lack of niacin in the diet causes pellagra, a debilitating disease associated with protein-deficient diets that is caused by insufficient niacin and tryptophan.

Third, alkaline cooking also reduces problems associated with mycotoxins in corn. Mycotoxins are produced by molds that grow on the kernels and cobs and can be very dangerous to both people and animals. Research shows that nixtamalization reduces the toxicity of these mycotoxins, although it does not completely eliminate the problem. Care should be taken to keep moldy corn out of the soup pot.

Finally, when alkaline solutions that contain calcium hydroxide (i.e., lime) are used, the calcium content of corn may be increased as much as 400 percent, greatly improving the quality of diets in which dairy products are absent or used sparingly.

Corn and Human Nutrition

Proteins, composed of amino acids, are essential for human health and are used by the body in transport, structure, and regulatory functions. Plants produce all the amino acids that they require, but human cells can synthesize only eleven of the twenty required amino acids. We get the other nine, called essential amino acids, from foods that already contain them. Protein content of corn usually varies from 8–11 percent of the kernel weight, making it a modest source of protein for human diets. Of the nine essential amino acids, corn provides sufficient amounts of the sulfur-containing amino acids but is deficient in lysine and tryptophan.

In contrast, beans and other legumes have large amounts of lysine and tryptophan but less of the sulfur amino acids. By eating both corn and beans, people are able to obtain all the essential amino acids. People who eat only corn with no legumes or other source of protein and do not treat the corn with calcium hydroxide are most at risk for pellagra. Native peoples in the Western Hemisphere who eat traditional foods have never suffered from pellagra because they prepare corn with an alkaline solution and eat beans along with corn. Even if meat or dairy products are not available, a diet based on corn and beans provides high-quality protein in sufficient quantities to meet human needs. It was only when corn was transported to other parts of the world (southeastern United States, Europe, and Africa, in particular) that pellagra appeared because neither the nixtamalization process nor the beans accompanied the corn.

Corn and Healthy Diets

Many Native Americans today suffer from diseases that may be closely linked to the foods they eat. Heart disease, diabetes, strokes, and some cancers are associated with diets that are high in refined carbohydrates (sugar and white flour) and animal-based fats. In some Native American communities, the incidence of diabetes is at epidemic levels, even though this disease was almost unknown among indigenous populations in North America before 1960. Many people hypothesize that obesity and diet-linked diseases flourished when Native peoples stopped eating traditional foods, such as corn, beans, and squash, and increased their consumption of white flour and other processed foods. Consequently, many health professionals in Native communities advocate a return to traditional foods, such as whole grain corn and beans, to combat diabetes and cardiovascular diseases.

Processing and Preparing Corn

Regardless of how the corn is prepared, cooks must vigilantly discard any discolored kernels before they begin preparing the corn because these kernels may be infected with mycotoxins, which can be highly poisonous.

Iroquois people are renowned for creating a multitude of savory dishes using corn (see figures 14 and 15). Many traditional

Figure 14 *Woman Preparing Corn,* painting by Ernest Smith. *Published with permission from the collections of the Rochester Museum & Science Center, Rochester, NY.*

dishes are made from immature or "green" corn, which is picked when the kernels are still milky. Arthur Parker's and F. W. Waugh's monographs provide information on these dishes.

Figure 15 *Woman Washing Corn*, painting by Ernest Smith. *Published with permission from the collections of the Rochester Museum & Science Center, Rochester, NY.*

Once the kernels are hard and dry, cooks have many additional options. The corn can be ground and used immediately in breads or porridges. Traditionally, Iroquois people used a mortar and pestle to grind the corn, but a hand grain mill can also be used (See figures 16 and 17, and sidebar 3, pages 44–45). Grinding corn decreases cooking time, and adding spices, beans, dried fruit,

Figures 16 and 17 **Mortar Construction.** *Courtesy of Robert F. Burt.*

and other sweeteners greatly increases the cook's repertoire. The corn can also be roasted first, which adds a subtle dimension to the corn's flavor. The References section contains information on sources for recipes.

Corn is also eaten as whole cooked kernels, which look more like cooked dry beans to those unfamiliar with corn in this form. These swollen, tender kernels serve as the foundation for many soups and stews. Corn is first boiled with wood ash (or baking soda or lye), using one part wood ash to one part corn. After cooking for about sixty minutes, the corn is rinsed under cold water, and the loosened hulls are removed by hand. The corn is usually boiled and rinsed several more times until the kernels are soft and all hulls are removed. The kernels will often change color as the cooking progresses. Tuscarora white flour corn starts out white, changes to bright yellow, and then after further cooking, becomes milky white.

Corn cooked this way is called hominy or posole. Either term refers to whole kernel corn that has been cooked in an alkaline solution until the hulls (pericarps) are removed and the corn is soft. You can also buy canned hominy in most grocery stores. A traditional Iroquois corn soup includes kidney beans and meat (i.e., salt pork, venison, or bear) that are added in the final stages of cooking (see sidebar 4, page 46). Whole kernel corn can also be roasted before it's made into hominy, giving the final product a more complex flavor.

After its initial preparation into hominy, hominy corn can also be ground (rather than used as a whole kernel) to provide additional textures and forms for even more recipes. For example, ground hominy corn is used to make tortillas and tamales in the Southwest and Mexico.

Sidebar 3

Constructing a Mortar and Pestle

You can construct your own mortar and pestle similar to those used by the Iroquois. It is challenging but can result in a useful tool that children (and adults) enjoy using to grind corn kernels. Robert Burt, retired staff member with Cornell's American Indian Program, constructed one in 2001 relying on photographs and descriptions from Arthur Parker's monograph *Iroquois Uses of Maize and Other Food Plants* (see the References section) as a guide.

Mortar: Burt started with a block of beech (*Fagus grandifolia*) measuring twenty-four inches high and eighteen inches in diameter. He used fire from charcoal briquettes to remove the wood, forming the bowl with successive burning and scraping. According to Parker, the Iroquois used pepperridge (*Nyssa sylvatica*) and black oak (*Quercus velutina*) to make their mortars. Choose a wood that does not split easily, such as sugar maple (*Acer saccharum*), sycamore (*Platanus occidentalis*), and yellow birch (*Fraxinus americana*). You can further inhibit cracking by drying the wood thoroughly before starting. The bowl of the completed mortar was thirteen inches deep and thirteen inches across its diameter. Here are the steps:

1. First, get a circular band of sheet steel four inches high, one-sixteenth inch thick, and about two inches less than the circumference of the block of wood. Place the band on top of the block in order to contain the charcoal.
2. Place charcoal briquettes within the band in a layer one or two briquettes deep and ignite them. You will probably need to increase the intensity of the fire by directing air onto the burning charcoal with

handheld bellows.

3. After ten minutes of burning, the surface of the block will be charred and glowing red-orange in color. Remove the briquettes and place them in a metal basket. Scrape the charred wood to remove it. (A small kit of scrapers with eight-inch handles and varying blade shapes is available from most hardware stores.)
4. Return the charcoal briquettes to the partially burned mortar, adding additional briquettes to maintain a small but intense fire. Repeat the burning and scraping, deepening the bowl with each successive scraping. You can deepen the bowl with alternate burning and scraping at a rate of about two inches per hour.
5. After the bowl is several inches deep, you may notice that the charcoal is burning the sides of the mortar, which is undesirable. Smear wet clay on the sides to prevent this and keep the soil wet with frequent applications of water from a spray bottle.
6. After the bowl is as deep as you want, smooth the interior surface using an angle grinder fitted with a four-inch sculpting wheel (available from woodworkers' supply stores or catalogs).
7. The final scraped surface can be blackened to give it an authentic charred surface by burning it lightly with a propane torch.

Pestle: Burt also used Parker's book to determine the measurements of the pestle. He started with a five-inch diameter sugar maple trunk and shaped it with a drawknife. The finished pestle was forty-five inches long and four inches in diameter. Again, start with dry wood, as the pestle will also crack if it is shaped while wet. According to Parker, a double-ended pestle adds desirable weight as well as providing two grinding surfaces.

Sidebar 4

Traditional Seneca Corn Soup Recipe

1 qt. hardwood ashes (important, softwood ashes will not work)

1 qt. white corn

½ lb. salt pork

½ c. red kidney beans or strawberry beans

Lg. cast-iron or enamel cooking kettle

Corn washing basket or colander

Fill a large kettle with about one gallon of water. (Ideally, this should be spring or well water, but plain old tap water is acceptable.) Boil water and add ashes. Lower heat to medium. Cook corn and ashes together at medium heat until the hulls of the corn come off (about an hour, time may vary depending on stove). Stir often, using a wooden paddle, so ashes and corn don't stick to the bottom. Also, be careful not to cook corn too high or too long, or it will disintegrate. Wash corn in the following way: Put corn in corn soup basket or colander and refill kettle with clean water. Place basket in kettle and swish around. This is to remove the hulls from the corn. Empty water and repeat two more times. Return corn to kettle and add fresh water. Lower heat and cook again for about an hour after corn starts to boil. In the meantime, soak beans and parboil. When the corn becomes transparent, and the heart of the corn comes out and is edible, wash again three times. Place corn, diced salt pork, and drained beans with about four quarts of water in kettle and cook until done, about three hours. Stir frequently and add water if necessary. This recipe makes about two-and-a-half quarts of soup.

Source: Lenore Abrams, Tonawanda Indian Reservation

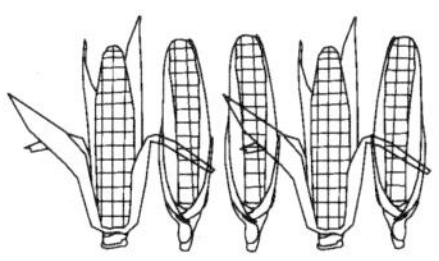

References & Sources of Information

Beginnings of Corn: A Haudenosaunee Perspective

Much of the information in this section comes from the translation and written expression of oral texts that are central to Haudenosaunee history. Sources used to write this book include:

Barnes, Barbara. *Traditional Teachings.* Cornwall Island, ON: North American Indian Travelling College, 1984. This book provides translations and transcriptions of the Creation Story, the Peacemaker and the Great Law by chiefs, faithkeepers, and elders of the Six Nations.

Gibson, John Arthur. *Concerning the League: The Iroquois League Tradition as Dictated by John Arthur Gibson in Onondaga.* Memoir 9. 1912. Translated and edited by Hanni Woodbury. Winnipeg, MB: Algonquian and Iroquoian Linguistics, 1992. In addition to a summary of the formation of the League by Woodbury based on Gibson's text, this book provides a line-by-line translation of Chief Gibson's recitation of the Great Law, originally transcribed in Onondaga.

Mann, Barbara. *Iroquoian Women: The Gantowisas.* New York: Peter Lang Publishing, 2000. Mann, a Seneca scholar, brings together in one place many of the sources regarding the role of Jigonsaseh and the corn growers in the Haudenosaunee history.

Parker, Arthur. "How the World Began." Chap. 1 in *Seneca Myths and Folktales.* Buffalo, NY: Buffalo Historical Society, 1923. This book contains oral stories collected and summarized by Arthur Parker, a Seneca ethnologist working in the early part of the twentieth century. This version of the Creation Story describes in some detail how various plants, including corn, arrived on Earth.

Parker, Arthur. *The Life of General Ely S. Parker*. Buffalo, NY: Buffalo Historical Society, 1912. In chapter four ("The Grand-Daughter of the Prophet"), Parker relates the story of Jigonsaseh, the corn cultivator who negotiated with the Peacemaker at the Confederacy's foundation.

Wallace, Paul. *White Roots of Peace*. 1946. Santa Fe, NM: Clear Light Publishers, 1994. Originally published in 1946, the reissued book contains illustrations by John Kahionhes Fadden, a foreword by Chief Leon Shenandoah, a message from Chief Sidney I. Hill, and an epilogue by John Mohawk. Although originally written by a nonnative scholar, this version of the foundation of the Iroquois Confederacy in the new edition of Wallace's original book has been contextualized and reaffirmed by contemporary Haudenosaunee chiefs, elders, and scholars. This version also describes Jigonsaseh's role at the Confederacy's foundation.

Corn Through Western Eyes

Fritz, Gayle J. "Multiple Pathways to Farming in Precontact Eastern North America." *Journal of World Prehistory* 4, no. 4 (1990): 387- 435. In this review article intended for academic audiences, Fritz provides an overview of western scholarship concerning the origins of agriculture in eastern North America. An extensive list of references provides access to much of the academic scholarship on the origins of agriculture in this part of the world.

Hart, John. "Maize, Matrilocality, Migration, and Northern Iroquoian Evolution." *Journal of Archaeological Method and Theory* 8, no. 2 (2001): 151-182. Hart, an ethnobotanist and archeologist working at the New York State Museum in Albany, has described much of the recent archaeological evidence for the appearance and distribution of corn in Iroquoia. Although he writes almost exclusively for academic audiences, this article provides an excellent summary of the most recent scholarship on the entry and distribution of corn into Iroquoia and further explores the role of women in the expansion of agriculture in the region. References from this article provide access to much of the academic scholarship on corn in the Eastern Woodlands.

Smith, Bruce D. *The Emergence of Agriculture*. New York: Scientific American Library, 1995. Smith provides in very accessible language an overview of the domestication of corn and its subsequent dispersal (with many of the details in dispute) throughout the Western Hemisphere. He also describes the domestication of plants such as *Chenopodium, Iva annua*, knotweed, and sunflower that were used by Native farmers in the Eastern Woodlands prior to the introduction of corn.

Thompson, Robert G., John P. Hart, Hetty Jo Brumbach, Robert Lusteck. "Phytolith Evidence for Twentieth-Century B.P. Maize in Northern Iroquoia." *Northeast Anthropology* 68 (2004): 25-40. This article provides information concerning the earliest dating of corn in New York.

Weatherwax, Paul. *Indian Corn in Old America*. New York: The Macmillan Company, 1954. Written more than half a century ago, Weatherwax describes with respect and enthusiasm the cultivation and use of corn among indigenous peoples across the Western Hemisphere. It also has wonderful photographs.

Corn at Contact

Cook, Frederick. *Journal of the Military Expedition of Major John Sullivan Against the Six Nations of Indians in 1779*. Auburn, NY: Frederick Cook. Knapp, Peck and Thompson, 1887. Written a century after the Sullivan expedition and filled with anti-Indian sentiment, this book is useful because of the detailed descriptions by Revolutionary War soldiers of agriculture in central Iroquoia in the late eighteenth century. These journals provide extraordinary snapshots that describe the scope and productivity of Iroquois corn growers.

Doolittle, William E. *Cultivated Landscapes of Native North America*. New York: Oxford University Press, 2000. Written by a geographer, this book is a compilation of documentary, ethnographic, and archaeological evidence on indigenous agriculture in North America. It is most useful as a means of finding the original descriptions of agriculture by European explorers/conquerors. Although the structure and arrangement of chapters is somewhat odd (a geographer's perspective on farming), the exhaustive list of references and many summary tables are very helpful.

Fussel, Betty. *The Story of Corn*. New York: Alfred A. Knopf, 1992. This book takes a hemispheric approach to corn, from its beginnings in Central America through its role in multiple indigenous civilizations and its impact within the United States. Reflecting the author's expertise, it includes much information on foods prepared from corn.

Hardeman, Nicholas P. *Shucks, Shocks and Hominy Blocks: Corn as a Way of Life in Pioneer America*. Baton Rouge, LA: Louisiana State University Press, 1981. This is a very readable, entertaining, and informative account of the role of corn in colonial life.

Parker, Arthur. "Iroquois Uses of Maize and Other Food Plants" in *Parker on the Iroquois*. 1913. Syracuse: NY, Syracuse University Press, 1968. Parker, a Seneca ethnologist working at the New York State Museum in Albany in the early part of the twentieth century, wrote one of the most authoritative and complete monographs on the use of corn in Iroquois communities from contact through the early twentieth century. He provides an excellent review of European accounts of the cultivation of corn in the Northeast and along the Atlantic Coast and provides detailed descriptions of the varieties and methods of cultivation, as well as ways of preparation. This book also has excellent photographs.

Walden, Howard. *Native Inheritance: The Story of Corn in America*. New York: Harper and Row, 1966. Although he has a dismissive attitude toward American Indian farmers, Walden provides an overview of the importance of corn in U.S. agriculture and industry.

Waugh, F. W. *Iroquois Foods and Food Preparation*. 1916. Ottawa: National Museums of Canada (National Museum of Man),

1973. Waugh, a contemporary of Parker, also produced a monograph on Iroquois crops with color photographs, descriptions of corn varieties in use at the time, traditional methods of the cultivation and harvest of corn, and food preparation.

Botany of Corn

Poehlman, John M. and David A. Sleper. *Breeding Field Crops*. Ames, IA: Iowa State University Press, 1995. This textbook describes the reproductive aspects of corn and explains the differences between commercial corn hybrids and open-pollinated varieties.

Smith, C. Wayne, Javier Betrán, and E. C. A. Runge, eds. *Corn: Origin, History, Technology, and Production*. Hoboken, NJ: Wiley, 2004. Directed at academics and researchers, this book covers corn from all aspects, including its growth and development, kernel physiology, breeding, pest management, harvest, processing, and much more. It is intimidating for non-scholars but thorough.

Finding Seed

Johnny's Selected Seeds. 955 Benton Ave., Winslow, ME 04901. 1-877-564-6697. HTTP://JOHNNYSEEDS.COM. They sell several varieties of open-pollinated corn seed.

The National Plant Germplasm System, part of USDA, collects thousands of maize varieties. Small quantities of open-pollinated seed may be available for experimentation and research. Information is available on their Web site (HTTP://WWW.ARS-GRIN.GOV/NPGS/).

Seed Savers Exchange. 3094 North Winn Road, Decorah, IA, 52101. Ph: (563) 382-5990. Fax: (563) 382-6511. HTTP://WWW.SEEDSAVERS.ORG. This is a nonprofit organization dedicated to the preservation of heirloom seeds. You can find several varieties of open pollinated corn.

Growing Corn

Almaco. 99M Avenue, Nevada, IA 50201-1558. (515) 382-3506. HTTP://ALMACO.COM. Almaco sells corn hand jabbers.

Bergstrom, G. C., W. J. Cox, G. A. Ferguson, S. D. Klausner, W. D. Pardee, W. S. Reid, R. R. Seaney, E. J. Shields, K. K. Waldron, and M. J. Wright. *Cornell Field Crops and Soils Handbook*. Ithaca, NY: Cornell Cooperative Extension, 1987. This is an excellent resource on New York soils and their management; it also provides excellent information on all aspects of crop production. It can be obtained from the Department of Crop and Soil Sciences Extension Office, 237 Emerson Hall, Cornell University, Ithaca, NY 14853. (607) 255-2177.

Bowman, Greg, ed. *Steel in the Field: A Farmer's Guide to Weed Management Tools*. Beltsville, MD: Sustainable Agriculture Network, 1997. This book provides practical information on how to control weeds with cultivators and other tools, with little or no use of herbicides. It has excellent information on how to select the appropriate weed control equipment for corn and other crops.

Cox, Bill, and Larissa Smith, eds. *2009 Cornell Guide for Integrated Field Crop Management*. Ithaca, NY: Cornell Cooperative Extension, August 2008. A practical guide for New York farmers growing field crops, it provides excellent information on all aspects of corn production. This publication is available online at HTTP://IPMGUIDELINES.ORG/FIELDCROPS/. A print version and other publications can also be purchased at this Web site.

Johnny's Selected Seeds. 955 Benton Ave., Winslow, ME 04901. 1-877-564-6697. HTTP://JOHNNYSSEEDS.COM. In addition to corn seed, you can purchase hoes and corn jabbers.

Lehman's. One Lehman Circle, Kidron, OH 44636. 1-888-438-5346. HTTP://WWW.LEHMANS.COM. You can find hand corn jabbers here.

Magdoff, Fred, and Harold van Es. *Building Soils for Better Crops*. Beltsville, MD: Sustainable Agriculture Network, 2000. This is an excellent reference for managing soils, whether you are using organic methods or more conventional strategies. It includes information on reducing soil erosion, improving drainage, and using cover crops, composts, and manures to improve soil fertility.

Managing Cover Crops Profitably. 2nd Ed. Beltsville, MD: Sustainable Agriculture Network, 1998. This is another excellent reference that explains how and why cover crops work and provides all the information needed to build cover crops into any farming operation.

National Sustainable Agriculture Information Service/Appropriate Technology Transfer for Rural Areas (ATTRA). Their Web site (HTTP://WWW.ATTRA.ORG/) provides access to a large body of information on organic and sustainable production practices. The articles are often free if you have access to the Internet or available for a low price if you order hard copies. Articles of particular interest include:

- "Organic Field Corn Production" by George Kuepper (January 2002).
- "Principles of Sustainable Weed Management for Croplands" by Preston Sullivan (2003).

Shedd, C. K. *Storage of Ear Corn on the Farm in the North Central States*. 1949. Farmers' Bulletin No. 2076. Washington, DC: United States Department of Agriculture, 1955. Written more than fifty years ago, this bulletin describes all the essential information for building and using corncribs.

Smith, C. Wayne. *Crop Production Evolution, History, and Technology*. New York: John Wiley and Sons, Inc., 1995. This textbook provides technical information for farmers on all aspects of growing corn, including planting, fertilization, weed control, harvest, and storage.

Sustainable Agriculture Research and Education (SARE). This program, part of USDA's Cooperative State Research, Education, and Extension Service, funds projects and conducts outreach designed to improve agricultural systems. Their Web site (HTTP://WWW.SARE.ORG/) provides access to many resources for developing farming systems that are profitable, environmentally sound, and good for communities. You can order any of the other Sustainable Agricultural Network books that are listed in this References section from the SARE Web site.

Eating Corn

Cookin' With Three Sisters. Available at the Mantaka American Indian Council Web site (HTTP://WWW.MANATAKA.ORG/PAGE175.HTML), this online recipe book, courtesy of the Oneida Nation, contains many recipes using corn, beans, and squash.

Jackson, Y. M. "Nutrition in American Indian health: Past, present, and future." *Journal of The American Dietetic Association* 86, no. 11 (1986): 1561-1565. This article provides an overview of the critical health issues of Native Americans and the impact of diet and nutrition on health.

Katz, S. H., M. L. Hediger, and L. A. Valleroy. "Traditional Maize Processing Techniques in the New World." *Science* 184, no. 4138 (1974): 765-773. This article provides an excellent description of nutritional aspects of corn with a clear description of nixtamalization, its history, and use by indigenous peoples around the world.

Maize in Human Nutrition. Rome: Food and Agriculture Organization of the United Nations, 1992. This publication, available online at HTTP://WWW.FAO.ORG/DOCREP, provides an excellent overview of the nutritional aspects of corn for human consumption, including the chemical composition of the kernel, the place of maize in human diets, and the effects of processing. It also includes an extensive bibliography.

Pleasant, Barbara. "Uncommon Corn." *Mother Earth News* (April/May 2004). This publication is available online at HTTP://WWW.MOTHEREARTHNEWS.COM/ORGANIC-GARDENING/2004-04-01/UNCOMMON-CORN.ASPX. Pleasant describes in nontechnical language how to grow and process corn. The article includes information on nutrition and the value of nixtamalization, and it provides encouragement and good advice for anyone who wants to grow, process, and eat their own corn.

White, Pamela J., and Lawrence A. Johnson, eds. *Corn: Chemistry and Technology*. St. Paul, MN: AACC International, 2003. Another academic reference for all aspects of corn, it is particularly strong on corn processing.

Education and Curriculum

Cornelius, Carol. *Iroquois Corn in a Culture-based Curriculum: a Framework for Respectfully Teaching about Culture.* Albany: State University of New York Press, 1999. Written by an Oneida scholar and educator, this book provides a theoretical framework, as well as specific strategies and practices, for incorporating traditional knowledge about corn into formal educational settings.

Eames-Sheavly, Marcia. T*he Three Sisters: Exploring an Iroquois Garden.* Ithaca, NY: Cornell Cooperative Extension, 1993. This publication is available online at HTTP://WWW.HORT.CORNELL.EDU/GBL/PUBS/INDEX.HTML. This excellent curriculum provides activities and resources for using Iroquois agriculture to teach science, social studies, and language arts to K–8 students.

The Iroquois Studies Association maintains a Web site (HTTP://WWW.OTSININGO.COM/), which provides access to an array of resources and information on the Iroquois Confederacy and the individual nations. It also has an extensive list of useful Web sites.

The New York State Museum Web site (HTTP://WWW.NYSM.NYSED.GOV/IROQUOISVILLAGE/) contains resources for teachers on many aspects of Iroquois history and culture, including a section on the Three Sisters.

Williammee, Janet, and Dave Rickard. *Feeding Body and Soul: Haudenosaunee Agriculture in the 19th Century.* Cooperstown, NY: The Farmers' Museum/New York State Historical Association, 2002. This teachers' guide enables students to explore the role of the Three Sisters in Haudenosaunee communities using primary and secondary sources, as well as hands-on-activities including gardening.

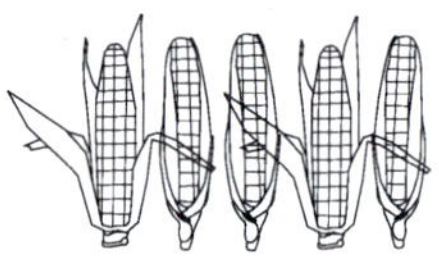

About NRAES

NRAES, the Natural Resource, Agriculture, and Engineering Service, is a not-for-profit program dedicated to assisting land grant university faculty and others in increasing the public availability of research- and experience-based knowledge. NRAES is sponsored by eight land grant universities in the eastern United States. Administrative support is provided by Cornell University, the host university.

NRAES publishes practical books of interest to fruit and vegetable growers, landscapers, dairy and livestock producers, natural resource managers, SWCD (soil and water conservation district) staff, consumers, landowners, and professionals interested in agricultural waste management and composting. NRAES books are used in cooperative extension programs, in college courses, as management guides, and for self-directed learning.

NRAES member universities are:

University of Connecticut	University of New Hampshire
University of Delaware	Rutgers University
University of Maine	Cornell University
University of Maryland	West Virginia University

Contact NRAES for more information about membership.

NRAES
Cooperative Extension
PO Box 4557
Ithaca, New York 14852-4557

Phone: (607) 255-7654 • Fax: (607) 254-8770
E-mail: NRAES@CORNELL.EDU
Web site: WWW.NRAES.ORG